Cushion 1
クッション1

Sofa
ソファー

Cheese 1
チーズ 1

Cheese 2
チーズ 2

Icebag
氷のう

Milk Carton
牛乳パック

Cushion 2
クッション 2

Doily
花瓶敷き

Tablecloth 1
テーブルクロス 1

Volleyball
バレーボール

Ladder
はしご

Cake ケーキ

Steps
足踏み台

Dumbbell
ダンベル

Rubik's Cube
ルービックキューブ

Ravioli
ラビオリ

Pillowcase
まくらカバー

Garbage Bag
ごみ袋

Garment Cover
衣類カバー

CENTER BACK

CENTER AND MATCH THIS LINE OF DOTS

NO. 6616 McCALL'S MULTI-BLUE

SH

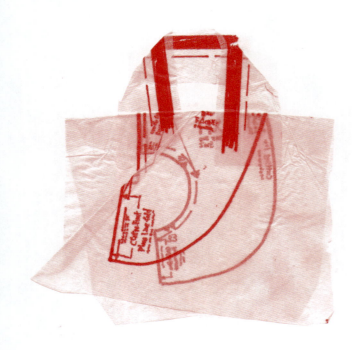

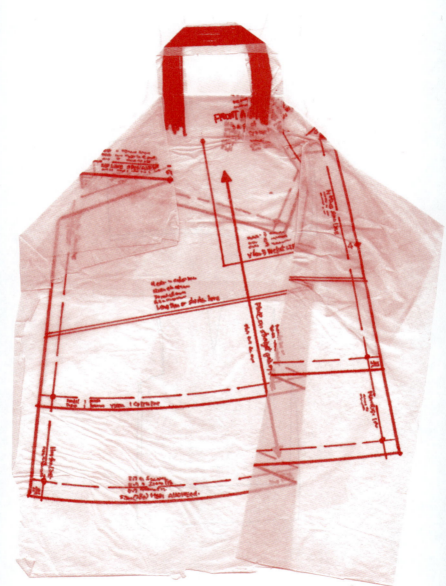

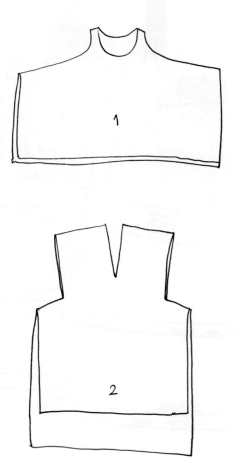

1

2

Garlic
にんにく

Toilet Paper
トイレットペーパー

Plastic Bag
レジ袋

Carpet
カーペット

Bed Sheets
ベッドシーツの服

T-shirt Poncho
Tシャツ

Shirt Poncho
シャツ

Poncho Dress
ワンピース

Poncho Coat
コート

Apron
エプロン

Bag
バッグ

Waist Porch
ウエストポーチ

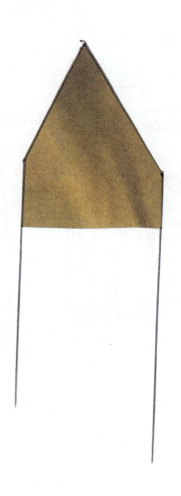

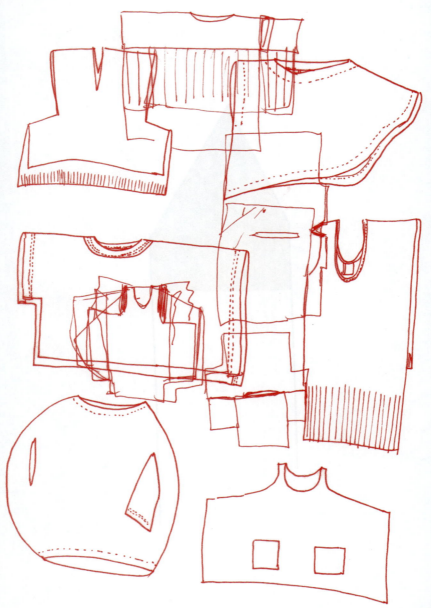

A. □01 □02

B. □01 □02

C. □01 □02 □03 □04 □05 □06 □07 □08
□09 □10 □11 □12 □13 □14

D. □01 □02 □03 □04 □05 □06 □07 □08
□09 □10 □11 □12 □13 □14 □15 □16
□17 □18 □19 □20

E. □01 □02 □03 □04 □05 □06 □07 □08
□09 □10 □11 □12 □13 □14 □15 □16
□17 □18 □19 □20

F. □01 □02 □03 □04 □05 □06 □07

G. □01 □02 □03 □04 □05 □06 □07

H. (

I. □01 □02 □03 □04 □05 □06 □07 □08
□09 □10 □11 □12 □13

J. □01 □02 □03 □04 □05

K. □01 □02 □03 (

L. (

M. (

N. □01 □02 □03

O. □01 □02 □03 □04 □05 □06 □07 □08
□09 □10 □11 □12 □13 □14 □15 □16
□17 (

P. □01 □02 □03 □04 □05 □06 □07 □08
□09 □10 □11 □12 □13 □14 □15 □16
□17 (

THERIACA

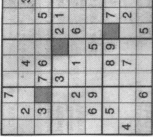
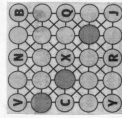

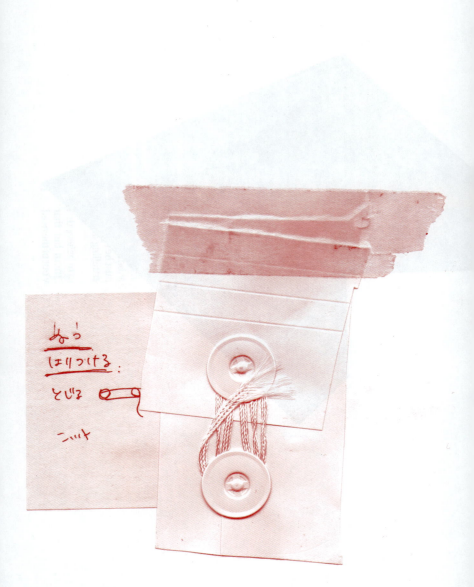

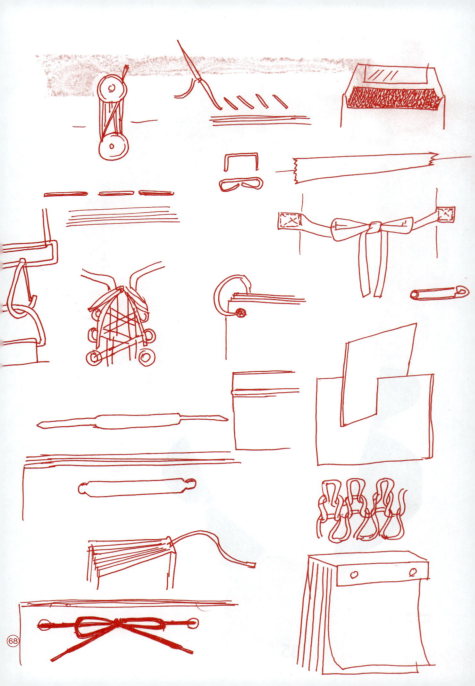

The limits of my geometry define the shape of my harlequin.

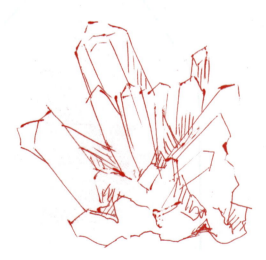

Shaping Clothes After Objects

Risa Hirota

This exhibition, entitled *Shapes and Forms: Clothes, The Body*, examines the echoes between the shape of a garment and the human figure, as well as the relationship between the two. Pieces of clothing shaped after objects, as opposed to templates known as *basic patterns*, act as facilitators in this process. It also happens to serve as the first large-scale exhibition by the fashion label THERIACA, operated by Berlin-based designer Asuka Hamada. One could consider basic patterns to be imaginary, fictional bodies; based on measurements of a human being's bust and height, these templates serve as the cornerstone of conventional garment-making. They allow one to create clothes that both adhere to and permit smooth movement of the body. The repeated processes of trial and error across the expansive history of clothing have refined and sophisticated this template, and for that reason, clothing manufacturers not only emphasize the template's importance but typically use the basic pattern in formulating their designs. The exhibition is constituted of three parts: a portion of works come from *Katachi no Fuku* (Shapes), which is a collection of garments rendered from geometric shapes. In the exhibition, we place special focus on the square-shaped pieces from that series. Along with these creations, we find garments from *Bed Sheets no Fuku* (Bed Sheets), which Hamada has conceived from large, square pieces of cloth. The third collection is *Mono no Katachi no Fuku* (Objects), featuring clothes shaped and molded after everyday objects. Each of the works departs from the confines of the basic pattern; and in reconstruing shapes with no inherent connection to clothing, one senses Hamada's efforts in discovering the potential found within such designs – designs that the basic pattern is incapable of creating. *Objects* is Hamada's most recent series and one that she created on the occasion of this exhibition. Her pieces in this group of works derive their shape from items that feel familiar and intimate to the designer herself – sofas, cushions, a Rubik's Cube, holed cheese, and more. One can discern semblances of these objects in her lineup: in the one-piece dress, crafted out of semi-transparent pink tulle (No. 25 p.5), or the one-piece colored the shade of café olé (No.29 p. 17) – frilled, elastic and adhering to the contours of the body. Her garments are ethereal, endearing, and yet eccentric all at once; they each possess a brightness and a sense of cheer, while also endowed with a magnetism that draws the eye. In this text, I would like to delve into thought on these new works while referring to Hamada's own experiences and her previous pieces.

One could consider Hamada's latest collection to be a continuation of her series, *Shapes*. The *Shapes* collection features geometric shapes – circles, squares, and crosses – rendered into pieces of clothing; Hamada first conceived the collection in 2013 in response to an assignment given to her during her time at the London College of Fashion. In the realm of garments, geometric shapes can function both as a base structure and as an embellishment. In pursuit of the essential question – *What is clothing?* – Hamada referred to a number of texts and began revisiting the history of clothes. In doing so, she came across the Egyptian kaftan and the Peruvian poncho: said to be the very first, original garments. Simplistic in structure, these pieces are composed of a single square of cloth with either one or three openings cut within (one for the head, or three for the head and both arms). At their essence, these pieces deviate from the practice of creating a garment with reference to the basic pattern – and what's more, the clothes bestow their wearers with an outline that is both unique and alluring. Hamada, whose education in both Japan and the U.K. had focused solely on the basic pattern as the cornerstone of clothes manufacturing, recalled feeling in her mind that the possibilities of clothing had expanded. She looked back on times and forms that existed before the advent of these constructs – an element easy to miss when one's focus has been trained on these models, forged after so many years of consideration – and decided to tap into them for inspiration in creating new clothing possibilities. As she referred to the original figures and structures of these pieces, she came to a realization that manifested in *Shapes* – a cloth of any shape can become a garment as long as this one, or these three, holes for the arms and/or head are present. Realized through experimental techniques, this collection gives way to works of ambition: ones that further push the envelope on what we define to be "clothes."

The following series, featuring pieces unfolding from large, square swaths of fabric, also came to fruition through the accumulation of these experiments. The group of pieces, collectively released here as *Bed Sheets*, features forms inspired by the poncho: a garment crafted from a single cloth with a hole for the head cut open at the top. One notices various differences in this large, square fabric depending on the location of this opening; the cloth forming the space around the hole seems to fall differently, while we also see changes in the drape that bestows a shape to the garment, together with the piece's outline and the sense of balance. The collection combines this component with shirts, one-pieces, waist pouches, and other contemporary, ready-made clothes, and we see the off-the-rack pieces accentuate the drape formed from the square cloth. The appearance of the clothes noticeably changes in sync with the wearer's movements. While at once delving into the potential of a single piece of fabric, one could also say this collection further deepens the dimensions of contemporary clothing design.

Hamada, when deliberating on something, places importance on first investigating that subject's most essential, fundamental state. One by one, she unearths the elements and conditions that have brought that object to being; once she reconsiders those components, she recombines them, reassembles them in different positions, and ponders creative possibilities. The process distances itself both from pre-existing notions and from Hamada's own conceptions. This technique slowly developed within the designer over time; it gives way to works of art while expanding upon the potential of clothing. The forms seen in THERIACA, the brand that she would come to establish afterward, serve as the products of such a technique.

THERIACA is the label Hamada founded during her days at the previously-mentioned London College of Fashion; titled after a miracle antidote in use from ancient times, the name embodies Hamada's desire to create clothes that help their wearers embrace a positive mindset. The notions of "fun" and "enjoyment" enliven people's expressions and brighten the air they have about them, and she places her faith in those forces and their universal power to bring people together.

Hamada's experience at her university in Japan helped foster such a mindset; it is an experience of crippling futility, together with the joy she felt in overcoming it. As part of a school assignment, Hamada had been in the process of creating an art piece out of textiles. She came to the realization, however, that as soon as her completed piece was taken down from the display, it would devolve into a simple object – nothing more. After all of the thought, time, and effort she had dedicated, all she would have to show for it would be an empty sentiment of self-satisfaction; she was left feeling resentful. However, the tides began to turn during her year-long study abroad session at a college of the arts in Canada. There, she studied textile design – of items made for human use, such as clothing and interior decoration – while exploring the possibilities of those forms. In doing so, she found that she was able to unearth a sense of purpose in her work for the first time. Hamada felt elated; she had found hope within these practical objects. These functional necessities allow for the involvement of a third party – a user – and Hamada found that this dynamic allowed her to share her own ideas and beliefs with her audience. She delighted in knowing that these methods allow for developments that transcend the imagination and came to realize that this element of uncertainty was precisely what she had been seeking. The experience taught her to not one-sidedly force her aesthetic as a designer onto an audience; she came to understand that a garment reaches completion through the intervention of a wearer – and that both wearers and viewers can use clothes as tools for communication and in bolstering their own emotional well-being. In essence, Hamada awoke to the desire to create clothes that would become more than just objects. This has become her driving principle as she continues with her creative endeavors.

Let us shift our attention to the creations Hamada has released up into the present; out of her body of work, her three handiwork books are likely the most well-known.

The first among them is *Katachi no Fuku* (Uniquely Shaped Clothes), which depicts human models wearing garments constructed from geometric shapes (shapes not created with a wearer in mind). This is her first endeavor fashioning clothes

that deviate from the basic pattern. Her second, *Ookina Fuku wo Kiru, Chiisana Fuku wo Kiru* (Oversized Clothes and Fitted Clothes), depicts garments that all share the same essential shape but vary by both measurement and size. A single model wears each of the pieces, illustrating differences in the fit; in other words, it draws attention to the varying forms of looseness born from the spaces between the body and a piece of clothing. The book elaborates on the appeal found in all garments – whether they be a bit smaller, a bit bigger, or just the right size – while her perspective reminds us of the uniqueness found in the shapes of our own bodies. In contemporary society, we are accustomed to purchasing and wearing ready-made-clothes. Everyone looks at garments as M's and L's, as size sixes; people select their clothes through a standard of proportion derived from the basic pattern. But perhaps to do so means to fit our bodies and thoughts into a framework of presumption. A body shrouded in clothes does not reveal the wearer's individuality in the same way that their face does, but this was not always the case; all bodies are of course different and unique. In reading, one realizes that those of any body type – whether short, husky, or slender – can wear and enjoy clothes in a way that is unique to them.

Silhouette and size, covered in these two texts, function as fundamental elements to a garment's composition. Changes made to these elements go on to represent shifts to fashion trends themselves: ones that have colored and defined the generations. The "body-conscious" trend popular in the latter half of the 1980's ushered in the advent of a style of close-fitting garments, emphasizing the silhouette and the outline of the body. In the present day in 2018, an "off-shoulder" style has come to gain prominence; one opens up extra room in the area around the breadth of the shoulders and leaves up plenty of space between the garment and the body.

The third of Hamada's texts, *Piisuwaaku no Fuku* (Piece Work Clothing), represents an attempt to distance herself from basic-pattern-centered processes through the use of patchwork – but in a way differing from that of *Katachi no Fuku*. A fundamental method in the realm of handicrafts, patchwork refers to a technique in which one sews small patches of cloth together. Today, people mainly consider patchwork to be a method for adding embellishments onto the surface of a garment – many harbor doubts about its application in creating the structure of the garment itself. However, here we see Hamada utilize patchwork as the main technique in realizing the garment's fundamental shape; she effectively uses patchwork not only as a medium of embellishment but also, in some surprising spots, as an essential transition-maker on the garment. In doing so, she depicts the technique in a new, fresh light. Hamada also emboldens the impact of the piece's coloration through the use of numerous, single-color cloths. The pages boast color arrangements both rhythmical and dynamic, laden with the sensation of movement.

These three texts resist the ephemerality of trends – her first volume, released in 2015, remains on shelves at booksellers up to present day in 2018. As handiwork books, her pieces have garnered an unprecedented response from audiences; and with translated editions released in China, Korea, France, Germany, and more, we see that their appeal has resounded overseas as well. The books, Hamada says, came about from her desire to communicate the joy of clothing found from beyond of the confines of retail – and in that regard, we could say she succeeded splendidly. Many creators from outside of the world of sewing or fashion own copies of her books: something attributed to the combination of the books' affordable price and superior designs. The concept is solid, and its contents delve into fundamental questions surrounding the origins of clothing. The proposed designs for all three texts are diverse yet thoughtful, while commentary and tips on sewing techniques line the pages – essentially, they more than adequately fulfill all the requirements needed for one to consider a text to be a book of reference. The refined sense of balance present within could be seen as a signature unique to Hamada herself.

In exploring Hamada's work, one can recognize common traits among them: notably, her apparent fascination with an object's fundamental shape, as well as how she uses experimental processes to "paint over" preconceptions and presumptions in an entertaining way. One could say the same for this new objects-rendered-as-clothing collection; the idea of fashioning clothes from the shapes of sofas and of cakes is an essentially youthful one, easily approachable and comprehensible by all audiences. And again, the garments here break free from the confines of the basic pattern. As in *Shapes*, Hamada fashions garments from shapes that have been devised for purposes other than wearability. The forms she presents here have evolved from the two-dimensional to the three-dimensional – but one could hardly label this the only difference between them. In a way essentially different from the geometric forms in *Shapes*, items possess both meaning and necessity; it has an assigned function and a form created for the purpose of fulfilling that function. Within these works, Hamada takes an item – an amalgamation of form and function – and strips away the element of *form*. She then bestows the form with a new purpose (to function as a garment) and shifts the dynamic of form and function in that object. In turning our eyes to the pieces born from this process, we could see this as an attempt to unearth several elements: the most essential requirements needed to realize a work of clothing and the essential roles and functionalities latent within the shapes of a garment. The shape of an object, when it comes into contact with the shape of the body, gives way to a dissonance of sorts; it can manifest as an accumulation (or overabundance) of cloth that creates the surprising illusion of volume on a body part or can conversely appear as a distortion. The clothes she creates appear as complicated and unique embellishments – ones unable to rendered through designing clothes purely as *clothes* – and begin to take on new

roles. One could say that this imbues elements of unpredictability and fascination throughout her pieces.

Asuka Hamada was raised by her father, an architect, and her mother, an art instructor for children. In her youth, she recalls there being a number of art texts and architectural magazines around the house, together with a number of materials that her mother would use in her lessons. Her mother's art classes not only focused on rendering pictures by paints and pens but also implemented elements of arts-and-crafts. She would instruct students to create to their heart's content in wielding a variety of materials, such as wool and origami paper. Hamada describes how she views cloth and thread to function in the same way that paints do; she likens the process of creating a garment to fashioning a sculpture from cloth, adding and removing colors and lines as she moves along. She describes how she is able to fold and pinch the fabric and threads, and how those factors lend themselves to a more intriguing experience than simply drawing a picture on a flat surface. For someone with so much interest in the primordial forms of objects, this statement is very Hamada-like – but we could also infer that this quality may find its origins in her early childhood environment. The same could be said of her style: one in which she employs a variety of materials and pieces them together through a process of trial and error. Hamada states that the predictable fails to interest her. *Shapes* and this new collection delve into what conditions give way to a garment while revolving around the incongruities found between the shapes of clothes and the shape of the body. Hamada's recently-published fourth volume, *Amai Fuku* (Sweet Clothes) is another endeavor that takes an in-depth look into the basic methodology of sewing; she focuses her attention on collecting or layering pieces of fabric as a method of creating and managing volume: such as gather, tuck, darts, and frills. The text also goes into detail on how to create a certain body outline (how to increase the volume on which part of the body and to what degree) while showing that these components of volume can similarly serve as embellishments. *Sweetness* serves as a keyword of the text as a whole, making it easy to assume that the clothes within have cuter, adorable qualities; but within, we find that *methodology*, a topic both earnest and deep-reaching, pervades the book as its central theme. While keeping with her experimental nature, I expect that Hamada will continue to present us with works that persist in inspiring us.

Curator, Iwami Art Museum

ひとでないものから「かたち」をとる服

廣田理紗

　「服のかたち／体のかたち」と題したこの展覧会は、原型でないものから形をとった衣服を通じて、衣服と身体の間に生じる形のひびきあいや、関係について考える試みだ。また、ベルリンを拠点に活動を展開する濱田明日香が手がけるファッションレーベル「THERIACA（テリアカ）」初の大規模展覧会でもある。原型とは人の胸囲や身長の寸法から割り出す架空の体ともいえる型のことであり、一般的には服作りの出発点ともなる形のことである。長い衣服の歴史の中で試行錯誤が重ねられ、身体にフィットする動きやすい服を作るためのベースとして洗練されてきた。そのため通常の衣服製作の場では、これを展開するデザインが一般的であり、重視されている。展覧会は、幾何学的な形を服へと落とし込んだ「かたちの服」、その中から四角形に注目し、四角の大きな布から発想を展開した「ベッドシーツの服」、そして「もの」から形をとり制作した「もののかたちの服」の三部構成となっている。いずれの作品も原型から離れることから出発した作品であり、本来は衣服とは無縁な形を、服として捉え直すことで、原型からでは生み出すことのできないデザインの可能性を探る試みともいえよう。「もののかたちの服」は、濱田が本展覧会のために手がけた新作であり、ソファーやクッション、ルービックキューブ、穴あきチーズなど濱田にとって身近な「もの」から形がとられた実際の作品は、薄く透き通るピンク色のチュールでできたワンピース（No.25、p5）や、カフェオレ色でフリルがついた、しなやかに体に沿うワンピース（No.29、p17）などといったラインナップで、軽やかで愛らしく、同時に少し不思議さがある服ばかりだ。またいずれも楽しさや明るさもあり、人を惹きつける吸引力も兼ね備えている。本稿ではこの新作について、濱田のこれまでの作品や経歴を参照しながら考えてみたい。

　このたびの新作は、「かたちの服」に続くコレクションと位置付けることができる。「かたちの服」は円形や四角形、十字形など幾何学的な形を衣服とする試みで、ロンドン・カレッジ・オブ・ファッション在学中の2013年に、課題に答える形で考案された。幾何学の形が構造と装飾、両方の役割を果たす衣服である。濱田はこの時、「そもそも服とは何なのか」という問いを持ち、様々な書籍を参照し、衣服の歴史を見直したという。その中で、エジプトのカフタンやペルーのポンチョに出会う。それらは原初の衣服とも言われるもので、四角い布に一つないし三つの穴（頭、あるいは頭と両腕を出すための）をあけただけのシンプルな構造であり、原型から起こして作る衣服とは発想の根本がまるで違うものだった。そしてそれらはいずれも人が着用した時に、ユニークで魅力的なシルエットを実現していた。それまで日本でもイギリスにおいても、原型を出発点とする教育を受けてきた濱田はこの時、服の可能性が広がるような感覚を得たという。長い時を経た末に完成された服の概念やルールに気を取られていると見落としてしまうが、それらが生じる前の原初の姿に立ち返り考えることで、新たな服の可能性を見出せると思うに至った。そして、原初の衣服の形と構造を参考に、どんな形のものでも頭や腕を出す穴を一つから三つ開ければ着られるものになるのではないか、と発想したのが「かたちの服」である。このコレクションを契機に、濱田は衣服として捉えることのできる範囲を押し拡げるような意欲的な作品を、実験的な手法を用いて生み出してゆく。

　続けて制作され、四角形の大きな布から発想を展開した作品もまた、そうした実験の積み重ねの上に立つものだ。本展で「ベッドシーツの服」としてまとめて発表されることとなったこの一連の作品は、一枚の布に、頭を出すための穴を一つ開けただけの衣服、ポンチョから着想を得ている。四角い大きな布は、穴を開ける場所によって、その周辺に広がる布の下がり方、布が形作るドレープや全体のシルエット、バランスに大きな変化が生じる。それをシャツやワンピース、ウエストポーチなど、すでに一定の型が出来上がっている現代的なアイテムと合わせることで、四角い大きな布が成すドレープを際立たせ、着用者の動きに応じて生じる表情の変化をより鮮明に感じられる作品に仕上げたものが、このコレクションである。一枚の布の持つ可能性を探ると同時に、現代的な衣服のデザインの幅を拡張しようとする作品といえる。

　このように濱田は物事を考えるときに、その原初の姿を確認することに重きを置いている。それを成立させている要素や条件を一つ一つ洗い出し、見つめ直したら、その要素どうしを組み直したり、他のものに当てはめたりして、新たな姿を模索する。こうした作業が、既成概念や自身の思い込みから距離を取り、服の可能性を広げながら作品を成すためのテクニックとして、徐々に濱田に定着していった。この手法は、のちに立ち上げられるレーベル「THERIACA」での作品制作へと引き継がれることとなる。

ところで「THERIACA」とは、濱田が先述のロンドン・カレッジ・オブ・ファッション在学中に立ち上げたレーベルである。その名は古くからある万能解毒剤から引用したもので、服が着用者の心をポジティブに転換する助けとなることを願ってつけられた。「面白さ」や「楽しさ」が、人の表情や雰囲気を明るく和ませ、人と人とをつなぐ普遍的な力を持つことに期待を込めている。

こうした思いを抱くに至った背景には、濱田が日本の芸術大学で学んでいた時代に遭遇した強い虚しさと、それを覆す喜びの体験がある。濱田は当時、課されるままに繊維を素材とした美術作品の制作に取り組んでいたが、出来上がった作品は発表が終われればただの「もの」となってしまう事実に直面した。どれだけ考え尽くし、時間や気持ちをかけて制作しても、自己満足で終わる感覚が辛く、やるせなさが残ったという。転機となったのは在学中に一年間留学したカナダの芸術大学での時間だった。そこでは、服やインテリアなど、人が使うことを前提としたテキスタイルのデザインや、その可能性を研究した。その時間の中で初めて自分の制作物に意義を見出し、強い喜びを得られたという。濱田はそこで、実用的なものに希望を見出した。実用品には他者が関わる余地があり、それを通して自身の考えや思いを共有することができると考えた。また、時として想像もしていないような発展をみせる場合があることが楽しく感じられ、それこそが求めていたことだと自覚するに至った。この経験から、デザイナーの美学を一方的に押し付けるようなものではなく、人が関わることで完成する服、さらには、着用者やそれを見た人、手にした人の気持ちに作用したり、コミュニケーションのきっかけとなるなど、「単なるもの」として終わることのない服を目指すという、濱田の作品作りにおける指針が確立された。

ここで濱田がこれまで発表してきた作品に目を移したい。濱田の作品として最も多くの人が目にしているのが、三冊の手芸本だろう。

一冊目となる『かたちの服』は先述の通り、着るために用意されたのではない幾何学的な形を着てみるという内容だ。原型から離れた服作りが実践された最初の例である。二冊目の『大きな服を着る、小さな服を着る』は同じかたちの服を、寸法違いで作り、同じ人物が着用した時のフィット感、つまり体と服との間に生まれるゆとりの具合の違いに注目する内容である。少し小さい

服も、少し大きい服も、ちょうどいいサイズの服とはちがう魅力があることを示し、私たち自身の体の個性を思い起こす視点をも提供している。既製服を購入して着ることが当たり前になっている現代においては、だれもが、自分はMだとか、Lだとか、36号だなどといって、原型を基準に整理されたサイズで服を選んでいる状況にあるが、それは思い込みの枠に自らの体と思考とをはめ込む行為なのかもしれない。いつも服に覆われている体は顔ほどにはその個性を意識されないが、本来はみな当然違い、身長の低い人は低いなりの、痩せた人も太った人もその体型にしか楽しめない服との付き合い方があることに気づかされる。

この二冊が取り扱ったシルエットやサイズといった問題は、衣服を構成する基本的な要件であり、その変化はそのまま流行の変化となってさまざまな時代を彩ってきた。1980年代後半に流行したボディーコンシャスは、体に密着して体の輪郭線を強調するスタイルであったし、2018年現在、街に溢れているのはオフショルダーと呼ばれる、肩幅を大きくとり、体と服との間にたっぷりとゆとりを取るスタイルである。

三冊目となる『ピースワークの服』は、いわゆるパッチワークの技法を使い、「かたちの服」とは違うアプローチで原型による服作りから離れる試みだ。パッチワークは小さな布を接ぎ繋ぐ、手芸の基本的な技法の一つだが、今日では主に表面を装飾する目的で用いられる技法と捉えられ、服の構造を作る技法とは考えにくくなっている。しかし濱田はそれを、服の形を成す中心的な技法として取扱い、装飾性を兼ねた意外な場所での「切り替え」として効果的に用いることで、この技法に対する新鮮なイメージを提供した。作品制作にあたっては単色の布を多用することで、色面構成の印象を強め、綺麗な色が並ぶリズミカルで躍動的な誌面づくりがなされている。

これら三冊は一時的な流行を追ったものとして消費されることなく、2015年に発売された一冊目が2018年現在においても書店に置かれ続けるなど、手芸本としては異例の反響を見せている。また、中国、韓国、フランス、ドイツなどで翻訳版が発行されるなどし、その影響は海外にまで及んでいる。服そのものの販売とは異なる手段で「服の面白さ」を伝えたい、という思いから出発したというこの試みは、大成功といえるだろう。手ごろな価格設定と、本のデザインの良さも相まって、濱田の本は、裁縫をしない人やファッションとは異なる分野のクリエイターたちも手にしているそうだ。しっかりとしたコンセプトがあり、衣服にまつわ

る根源的な問いを抱えた内容であるが、三冊ともで着まわしのき
くシンプルな、しかし気の利いたデザインが提案され、裁縫のコ
ツなどのコメントもあって、実用書としての要件もうまく満たし
ている。ここにみられるこうしたバランス感覚の良さも、濱田の
特質といえるだろう。

　以上に見てきたように、濱田の作品には共通して、物事の原始
的な姿への関心と、実験的な制作スタイルによって既成概念や思
い込みを「楽しく」塗り替えてしまうという特徴がみとめられる。
この特徴は「もの」から形をとった新作についても、同様に指摘
できるだろう。新作にみられるソファーやケーキを服にしてみよ
うなどという発想は、子どものそれのようで、誰にとっても親し
みやすく、わかりやすい。また新作は、原型によらない服であ
り、着るためではない形を衣服として捉え直す試みだという点で
「かたちの服」と共通している。しかしその変化は、単に取り扱
う形が二次元のものから三次元のものになった、というだけにと
どまらないだろう。「もの」の形には、「かたちの服」で取り扱っ
た幾何学的な形とは異なり、意味や必然性を有している。あらゆ
る「もの」には用途があり、「もの」の形状はその用途に応じて備
わったものだ。濱田の新作は、用途と一体となって「もの」を成
立させているその形を引き剥がし、別の（衣服としての）用途を
与えることで、形と用途をずらす。今回の試みは、そうして成立
した服をとおして、衣服を成立させる要件や、一般的な衣服の形
状に宿る役割や機能性をあぶり出そうとするものとみることもで
きる。「もの」の形は、体の形と出会った時にある種の不調和を
起こし、それが布の重なり（余り）や意外な部位でのボリューム、
あるいは歪みとなって現れている。そして、衣服として設計され
たものには出すことのできない複雑でユニークな装飾となって新
たな役割を獲得し、衣服全体に意外性や面白さを付与しているだ
ろう。

　濱田は建築家の父と、子どもたちに絵を教えている母の元で育
った。美術書や建築雑誌が身近にあったことはもちろん、母の教
室で使うための様々な素材が家にはたくさんあったという。絵の
教室といっても絵の具やペンで絵を描くというだけでなく、造形
教室といった雰囲気もあったようで、毛糸や折り紙など様々な材
料を使って、思い思いのイメージを形にする指導がなされていた
らしい。濱田は、布や糸を絵の具のようなものとして捉えている
という。濱田にとって服を作る行為は、色や線を足し引きしなが
ら布の彫刻を作っているような感覚の行為であり、布や糸は折っ
たりつまんだりできる点で、平面に描く絵より興味深いものだと
いう。物事の原初的な姿に関心を置く濱田らしい発言だが、その
出発点には濱田の生育環境が影響しているのかもしれない。様々
な素材を用いてあれやこれやと試行錯誤するスタイルもまた然り
であり、結果が予想できることには関心が持てないともいう。
「かたちの服」や新作は服の形と体の形のずれから、服の成立要
件について考える研究だった。先日刊行されたばかりの四冊目の
手芸本『甘い服』も、裁縫の基本的な技法と向き合った一冊とな
っている。ギャザーやタック、ダーツ、フリルという、布を寄せ
たり重ねたりしてボリュームを調整する技法に注目し、それをど
の部位に、どの分量使うかによって、シルエットを構築する手段
にも、装飾的ディテールにもなることを示す内容である。「甘さ」
を全体のキーワードとしているため、服の雰囲気がテーマのか
わいい本のように思われるかもしれないが、実際にはやはり技法
研究という真面目で奥深いテーマが横たわっている。濱田のこう
した実験的態度は今後も続き、私たちを驚かせる作品を次々と提
案してゆくのだろう。

（島根県立石見美術館主任学芸員）

Acknowledgements

We would like to express our hearty gratitude to the following whose assistance has greatly helped the exhibition and this book.

Mizuki Kin
Kiyoko Kubota
Tomoko Koizumi
Kana Kobayashi
Sayuri Sakairi
Nozomi Sumikawa
Kaori Tanaka
Michie Tanaka
Mariko Nishitani
Miyoko Nomura
Kiyoko Hazeyama
Sachiko Hamada
Susumu Hamada
Naoko Higashi
Sui Hiyama
Kazuko Matsuoka
Yu Miyakoshi

Sascha Tyl
Miho Shida
Shin Lee

謝辞

本展覧会の開催および本書発刊にあたりご協力を賜りました皆さまに深い感謝の意を表します。

金瑞姫
久保田清子
小泉智子
小林香菜
坂入小百合
澄川希望
田中薫
田中道枝
西谷真理子
野村三代子
櫨山喜代子
濱田幸子
濱田将
東直子
緋山粋
松岡和子
宮越裕生

Sascha Tyl
Miho Shida
Shin Lee

（五十音順、敬称略）

THERIACA

The fashion label of Asuka Hamada, a Berlin-based fashion designer. Titled after a miracle antidote of the same name, Hamada hopes that her works serve as catalysts in further improving upon the quality of life of their wearers. The subject of steady acclaim and praise, her patterns transcend the confounds of traditional forms and give way to imaginative, new shapes. Her pieces have won over the hearts of many spectators: her style is light and graceful while jocose at the same time.

After studying textile design at the Kyoto City University of the Arts and Nova Scotia College of Art and Design (Canada), Asuka Hamada worked in apparel planning for several years before making her way to the United Kingdom. At the London College of Fashion, Hamada engaged in research on fashion design and pattern cutting – first founding her own label, THERIACA, while enrolled there in 2014. Hamada relocated to Berlin in 2015, further expanding her sphere of activity. Making use of experimental methods, Hamada places distance between herself and the concepts of trendiness, convenience, and traditional conceptions in the pursuit of the possibility latent in clothing. Not only does she exhibit her clothes in galleries and museums and sell them within retail outlets, but she has released a variety of works, by way of book and otherwise, utilizing a variety of methods. Her authored works are *"Katachi no Fuku,"* (Uniquely Shaped Clothes) *"Ookina Fuku wo Kiru, Chiisana Fuku wo Kiru,"* (Oversized Clothes and Fitted Clothes) *"Piisuwaaku no Fuku,"* (Piece Work Clothing) and *"Amai Fuku"* (Sweet Clothes), all released from Bunka Publishing Bureau.
www.theriaca.org

THERIACA（テリアカ）

ベルリンを拠点に活動を展開するファッションデザイナー、濱田明日香によるレーベル。万能解毒剤からとられた名前には、作品が人のコンディションをポジティブに転換する力となるようにとの願いが込められている。原型にとらわれないパターンが作り出すフォルムに定評があり、軽やかでありながらどこかユーモラスな作風は多くの人の心をとらえてきた。

濱田明日香は京都市立芸術大学、ノヴァスコシア芸術大学（カナダ）にてテキスタイルデザインを勉強した後、デザイナーとしてアパレル企画に数年携わり、渡英。ロンドン・カレッジ・オブ・ファッションにてファッションとパターンについて研究し、2014年、在学中に自身のレーベル "THERIACA" をスタートした。2015年からはベルリンに拠点を移し活動を展開。実験的な手法を用いて流行や利便性、固定観念から距離を取りながら服の可能性をさぐっている。ギャラリーや美術館での展示、ショップでの販売に加え、書籍などいろいろな方法で作品を発表してきた。著書に「かたちの服」「大きな服を着る、小さな服を着る」「ピースワークの服」「甘い服」（すべて文化出版局）がある。
www.theriaca.org

List of Works
作品リスト

作品番号 / Number
Title
Material
作品名
素材
ページ番号 / Pages

Note
The order in which the works are presented within the catalog is different from the composition of the actual exhibition in some sections. Also, note that there are exhibited works whose photographs have not been listed within the catalog.

実際の展示構成と
カタログ掲載順とは異なる部分がある。
また出品作品の一部に
写真掲載されていないものがある。

Shapes
かたちの服

1
Caterpillar
Acrylic Felt
いもむし型
アクリルフェルト
⑰

2
Scarecrow
Acrylic Felt
かかし形
アクリルフェルト
⑧⑤

3
T
Wool Felt
T型
ウールフェルト
⑨③

4
Bobhead
Wool Felt
おかっぱ型
ウールフェルト
⑧⑨

5
+
Wool Felt
十型
ウールフェルト
⑨⑥

6
Puddle
Acrylic Felt
水たまり型
アクリルフェルト
⑧④

7
Hexagon
Acrylic Felt
六角形
アクリルフェルト
⑨②

8
Worm-eaten Pentagon
Acrylic Felt
虫くい五角形
アクリルフェルト
⑧⑧

9
Square
Acrylic Felt
正方形
アクリルフェルト
⑧③

10
Keyhole
Acrylic Felt
かぎ穴型
アクリルフェルト
⑨①

11
Circle
Acrylic Felt
円
アクリルフェルト
⑧②

12
Upside Down T
Acrylic Felt
逆T型
アクリルフェルト
⑧⑦

13
Upper Half Circle
Acrylic Felt
上半円
アクリルフェルト
⑨⑤

14
Deformed Hexagon
Acrylic Felt
変形六角形
アクリルフェルト
⑨⓪

15
Lower Half Circle
Acrylic Felt
下半円
アクリルフェルト
⑧⑥

16
Rectangle
Acrylic Felt
長方形
アクリルフェルト
⑨④

17
Pie Chart
Acrylic Felt
円グラフ
アクリルフェルト
⑨⑧

Bed Sheets
ベッドシーツの服

18
T-shirt Poncho
Polyester Broadcloth
Tシャツ
ポリエステルブロード
⑪⑦

19
Shirt Poncho
Polyester Broadcloth
シャツ
ポリエステルブロード
⑪⑨

20
Poncho Coat
Polyester Broadcloth
コート
ポリエステルブロード
⑤③

21
Poncho Dress
Polyester Broadcloth
ワンピース
ポリエステルブロード
⑤①

22
Apron
Polyester Broadcloth
エプロン
ポリエステルブロード
⑤⑤

23
Waist Porch
Polyester Broadcloth
ウエストポーチ
ポリエステルブロード
⑤⑦

24
Bag
Polyester Broadcloth
バッグ
ポリエステルブロード
⑤⑥

Objects
もののかたちの服

25
Sofa
Polyester Tulle
ソファー
ポリエステルチュール
⑤

26
Rubik's Cube
Polypropylene Non - woven Fabric
ルービックキューブ
ポリプロピレン不織布
⑳

27
Cushion 1
Cotton Quilting
クッション 1
コットンキルティング
③

28
Dumbbell
Acrylic Felt
ダンベル
アクリルフェルト
⑲

29
Cake
Viscose Mock Leno
ケーキ
ヴィスコースモク・レノ
⑰

30
Cushion 2
Cotton Quilting
クッション 2
コットンキルティング
⑪

31
Doily
Polyester Gingham
花瓶敷き
ポリエステルギンガム
⑫

32
Milk Carton
Acrylic Felt
牛乳パック
アクリルフェルト
⑩

33
Toilet Paper
Viscose Crepe
トイレットペーパー
ヴィスコースクレープ
⑬

34
Ravioli
Wool Felt
ラビオリ
ウールフェルト
㉑

35
Ladder
Wool Felt
はしご
ウールフェルト
⑮

36
Volleyball
Polyester Chiffon
バレーボール
ポリエステルシフォン
⑭

37
Cheese 1
Polypropylene Non - woven Fabric
チーズ 1
ポリプロピレン不織布
⑦

38
Cheese 2
Polyester Mesh
チーズ 2
ポリエステルメッシュ
⑧

39
Pillowcase
Cotton Gingham
まくらカバー
コットンギンガム
㉓

40
Tablecloth 1
Polyester Chiffon Gingham
テーブルクロス 1
ポリエステルシフォンギンガム
⑬

41
Tablecloth 2
Polyester Gingham
テーブルクロス 2
ポリエステルギンガム

42
Garbage Bag
Nylon Parachute Cloth
ごみ袋
ナイロンパラシュート布
㉔

43
Garlic
Polyester Tulle
にんにく
ポリエステルチュール
㊷

44
Icebag
Cotton Twill
氷のう
コットンツイル
⑨

45
Bubble Wrap
Polyester Quilt Batting
エアーキャップ
ポリエステルキルト芯

46
Carpet
Polyester Canvas
カーペット
ポリエステル帆布
㊺

47
Plastic Bag
Polyester Mesh
レジ袋
ポリエステルメッシュ
㊹

48
Teabag
Polyester Jersey
ティーバッグ
ポリエステルジャージー

49
Garment Cover
Polypropylene Non - woven Fabric
衣類カバー
ポリプロピレン不織布
㉕

50
House
Acrylic Felt
家
アクリルフェルト
①

51
Steps
Cotton Jersey
足踏み台
コットンジャージー
⑱

Shapes
かたちの服

Circle
円

Square
正方形

Puddle
水たまり型

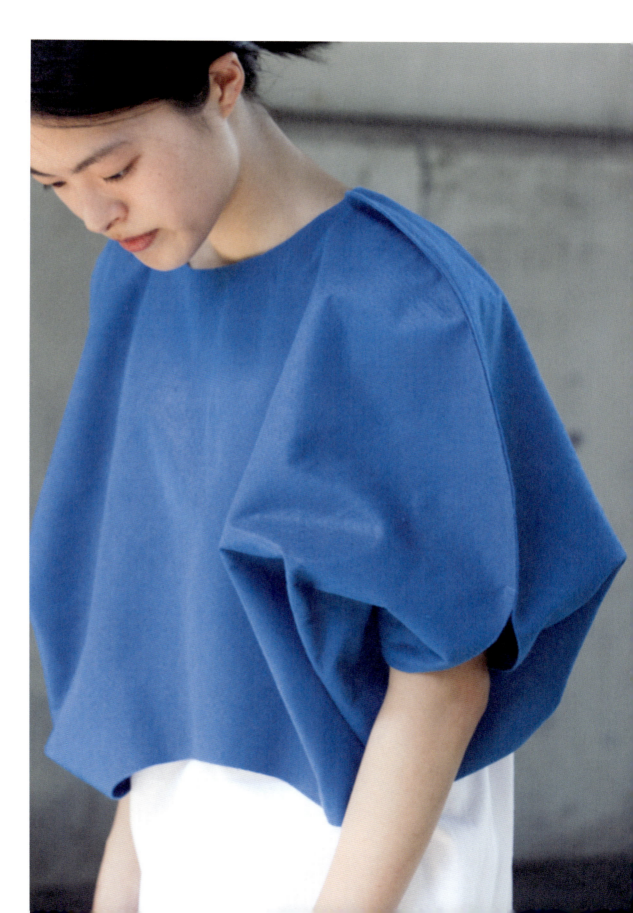

Scarecrow
かかし形

Lower Half Circle
下半円

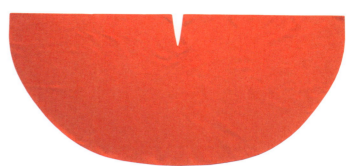

Upside Down T
逆T型

Worm-eaten Pentagon
虫くい五角形

Bobhead
おかっぱ型

Deformed Hexagon
変形六角形

Keyhole
かぎ穴型

Hexagon
六角形

T
T型

Rectangle
長方形

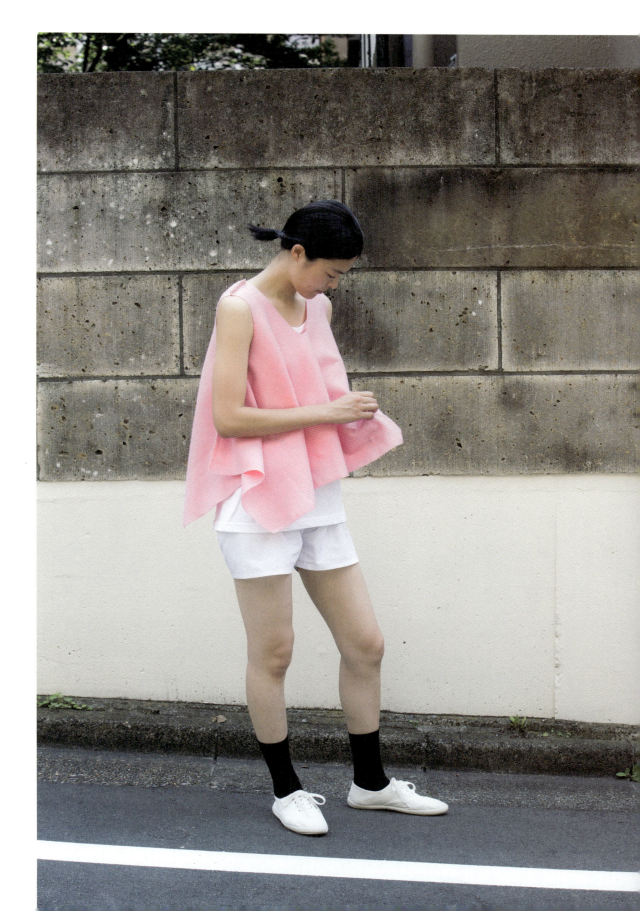

Upper Half Circle
上半円

十型

Caterpillar
いもむし型